MEMOIRS
of the
American Mathematical Society

ISSN 0065-9266 (print)
ISSN 1947-6221 (online)

Number 1332

AF620988

Corrigendum and Improvements to "Carleman Estimates, Observability Inequalities and Null Controllability for Interior Degenerate Nonsmooth Parabolic Equations" and Its Consequences

Genni Fragnelli
Dimitri Mugnai

July 2021 • Volume 272 • Number 1332 (second of 7 numbers)

AMS AMERICAN MATHEMATICAL SOCIETY

Library of Congress Cataloging-in-Publication Data

Cataloging-in-Publication Data has been applied for by the AMS.
See http://www.loc.gov/publish/cip/.
DOI: https://doi.org/10.1090/memo/1332

Memoirs of the American Mathematical Society

This journal is devoted entirely to research in pure and applied mathematics.

Subscription information. Beginning with the January 2010 issue, *Memoirs* is accessible from `www.ams.org/journals`. The 2021 subscription begins with volume 269 and consists of six mailings, each containing one or more numbers. Subscription prices for 2021 are as follows: for paper delivery, US\$1085 list, US\$868 institutional member; for electronic delivery, US\$955 list, US\$764 institutional member. Upon request, subscribers to paper delivery of this journal are also entitled to receive electronic delivery. If ordering the paper version, add US\$20 for delivery within the United States; US\$80 for outside the United States. Subscription renewals are subject to late fees. See `www.ams.org/help-faq` for more journal subscription information. Each number may be ordered separately; *please specify number* when ordering an individual number.

Back number information. For back issues see `www.ams.org/backvols`.

Subscriptions and orders should be addressed to the American Mathematical Society, P. O. Box 845904, Boston, MA 02284-5904 USA. *All orders must be accompanied by payment.* Other correspondence should be addressed to 201 Charles Street, Providence, RI 02904-2213 USA.

Copying and reprinting. Individual readers of this publication, and nonprofit libraries acting for them, are permitted to make fair use of the material, such as to copy select pages for use in teaching or research. Permission is granted to quote brief passages from this publication in reviews, provided the customary acknowledgment of the source is given.

Republication, systematic copying, or multiple reproduction of any material in this publication is permitted only under license from the American Mathematical Society. Requests for permission to reuse portions of AMS publication content are handled by the Copyright Clearance Center. For more information, please visit `www.ams.org/publications/pubpermissions`.

Send requests for translation rights and licensed reprints to `reprint-permission@ams.org`.

Excluded from these provisions is material for which the author holds copyright. In such cases, requests for permission to reuse or reprint material should be addressed directly to the author(s). Copyright ownership is indicated on the copyright page, or on the lower right-hand corner of the first page of each article within proceedings volumes.

Memoirs of the American Mathematical Society (ISSN 0065-9266 (print); 1947-6221 (online)) is published bimonthly (each volume consisting usually of more than one number) by the American Mathematical Society at 201 Charles Street, Providence, RI 02904-2213 USA. Periodicals postage paid at Providence, RI. Postmaster: Send address changes to Memoirs, American Mathematical Society, 201 Charles Street, Providence, RI 02904-2213 USA.

© 2021 by the American Mathematical Society. All rights reserved.

This publication is indexed in *Mathematical Reviews*®, *Zentralblatt MATH*, *Science Citation Index*®, *Science Citation Index*™*-Expanded*, *ISI Alerting Services*℠, *SciSearch*®, *Research Alert*®, *CompuMath Citation Index*®, *Current Contents*®*/Physical, Chemical & Earth Sciences.*
This publication is archived in *Portico* and *CLOCKSS*.
Printed in the United States of America.

∞ The paper used in this book is acid-free and falls within the guidelines established to ensure permanence and durability.
Visit the AMS home page at `https://www.ams.org/`

10 9 8 7 6 5 4 3 2 1   26 25 24 23 22 21

Contents

Abstract

This paper is a corrigendum of one hypothesis introduced in *Mem. Amer. Math. Soc.* **242** (2016), no. 1146, and used again in *J. Differential Equations* **260** (2016), pp. 1314–1371 and *Adv. Nonlinear Anal.* **6** (2017), pp. 61–84]. We give here the corrected proofs of the concerned results, improving most of them.

Received by the editor November 3, 2017.

Article electronically published on September 27, 2021.

DOI: https://doi.org/10.1090/memo/1332

2020 *Mathematics Subject Classification.* Primary 35K65, 93B05, 93B07.

Key words and phrases. Degenerate equation, singular equation, interior degeneracy, interior singularity, Carleman estimates, observability inequalities.

The first author is a member of the Gruppo Nazionale per l'Analisi Matematica, la Probabilità e le loro Applicazioni (GNAMPA) of the Istituto Nazionale di Alta Matematica (INdAM) and was supported by the GNAMPA project 2017 *Comportamento asintotico e controllo di equazioni di evoluzione non lineari.*

The second author was supported by the 2017 INdAM-GNAMPA Project *Equazioni Differenziali Non Lineari* and by the Italian MIUR project *Variational methods, with applications to problems in mathematical physics and geometry* (2015KB9WPT_009).

©2021 American Mathematical Society

CHAPTER 1

Introduction

In Chapter 5 of [**2**] there is a mistake in one assumption, which do not affect the results of the paper in the previous chapters. Precisely, the assumption in equation (5.2) has to be replaced by equation (5.6) therein. This replacement propagates in Hypotheses 5.2 and 5.3, as well as in Sections 4 and 5 of [**1**] and in Hypothesis 4.1(ii) of [**3**]. For the readers' convenience, we write here all the proofs of the involved results, though they are very similar. We remark that in all cases, replacing the correct assumptions makes all the written results true, and in most cases the proofs we give here are simpler than before. Each section follows the notations and the enumeration of the corresponding paper.

The authors apologize for any inconvenience caused.

CHAPTER 2

Correction to [2]

Hypothesis 5.1 must be replaced by

HYPOTHESIS 5.1. The subset ω is such that

- it is an interval which contains the degeneracy point:

$$\omega = (\alpha, \beta) \subset (0,1) \text{ is such that } x_0 \in \omega, \tag{2.1}$$

  or

- $\omega = \omega_1 \cup \omega_2$, where ω_i, $i = 1,2$, are intervals each of them lying on different sides of the degeneracy point, more precisely:

$$\omega_i = (\lambda_i, \beta_i) \subset (0,1),\ i = 1,2, \text{ and } \beta_1 < x_0 < \lambda_2. \tag{2.2}$$

REMARK 1. Observe that, if (2.1) holds, we can find two subintervals $\omega_1 = (\lambda_1, \beta_1) \subset\subset (\alpha, x_0), \omega_2 = (\lambda_2, \beta_2) \subset\subset (x_0, \beta)$.

As a consequence, Hypotheses 5.2 and 5.3 now read as follows:

HYPOTHESIS 5.2. Hypothesis 4.1 is satisfied. Moreover, if Hypothesis 1.1 holds, we assume that there exist two functions $\mathfrak{g} \in L^\infty_{\text{loc}}([-\beta_1, 1] \setminus \{x_0\})$, $\mathfrak{h} \in W^{1,\infty}_{\text{loc}}([-\beta_1, 1] \setminus \{x_0\}, L^\infty(0,1))$ and two constants $\mathfrak{g}_0$, $\mathfrak{h}_0 > 0$ such that $\mathfrak{g}(x) \ge \mathfrak{g}_0$ for a.e. x in $[-\beta_1, 1]$ and

$$-\frac{\tilde a'(x)}{2\sqrt{\tilde a(x)}}\left(\int_x^B \mathfrak{g}(t)dt + \mathfrak{h}_0\right) + \sqrt{\tilde a(x)}\mathfrak{g}(x) = \mathfrak{h}(x,B) \quad \text{for a.e. } x \in [-\beta_1, 1], B \in [0,1] \tag{2.3}$$

with $x < B < x_0$ or $x_0 < x < B$, where

$$\tilde a(x) := \begin{cases} a(x), & x \in [0,1], \\ a(-x), & x \in [-1,0]. \end{cases}$$

Here β_1 is the number appearing in (2.2) or in Remark 1.

HYPOTHESIS 5.3. Hypothesis 4.2 is satisfied. Moreover, if Hypothesis 1.1 holds, then there exist two functions $\mathfrak{g} \in L^\infty_{\text{loc}}([-\beta_1, 1] \setminus \{x_0\})$, $\mathfrak{h} \in W^{1,\infty}_{\text{loc}}([-\beta_1, 1] \setminus \{x_0\}, L^\infty(0,1))$ and two constants $\mathfrak{g}_0$, $\mathfrak{h}_0 > 0$ such that $\mathfrak{g}(x) \ge \mathfrak{g}_0$ for a.e. x in $[-\beta_1, 1]$ and

$$\frac{\tilde a'(x)}{2\sqrt{\tilde a(x)}}\left(\int_x^B \mathfrak{g}(t)dt + \mathfrak{h}_0\right) + \sqrt{\tilde a(x)}\mathfrak{g}(x) = \mathfrak{h}(x,B) \quad \text{for a.e. } x \in [-\beta_1, 1], B \in [0,1] \tag{2.4}$$

with $x < B < x_0$ or $x_0 < x < B$.

Using new Hypothesis 5.1, Corollaries 5.1 and 5.2 become part of Proposition 5.1 and Proposition 5.3, and Theorems 5.2 and 5.4 part of Theorems 5.1 and 5.3, respectively. Hence in subsections 5.1 and 5.2 we find only new Proposition 5.1, Theorem 5.1, Proposition 5.3 and Theorem 5.3 with the new hypotheses. As a consequence, in the statement of Proposition 6.2 condition (5.2) has to be replaced by (2) above. The new proofs, which are adaptations of those in [**2**], are presented below.

The divergence case

Proposition 5.1 (and consequent Theorem 5.1) has the same statement with new Hypotheses 5.1 and 5.2. Lemma 5.1, which is independent of (2.2), obviously holds, but the proof can be improved in the last part, starting from (5.20).

PROOF OF THE SECOND PART OF LEMMA 5.1. Take ω_1, ω_2 as in Remark 1. If v solves (5.7), define, also for later uses,

$$W(t,x) := \begin{cases} -v(t,-x), & x \in [-1,0], \\ v(t,x), & x \in [0,1], \\ -v(t,2-x), & x \in [1,2]. \end{cases} \tag{2.5}$$

Take $\rho : [-1,1] \to [0,1]$ as defined after (5.20) and set $Z = \rho W$. Then Z solves

$$\begin{cases} Z_t + (\tilde{a} Z_x)_x = \bar{h}, & (t,x) \in (0,T) \times (-\beta_1, \beta_1), \\ Z(t,-\beta_1) = Z(t,\beta_1) = Z_x(t,-\beta_1) = Z_x(t,\beta_1) = 0, & t \in (0,T), \end{cases} \tag{2.6}$$

with $\bar{h} = (\tilde{a}\rho_x W)_x + \tilde{a}\rho_x W_x$, supported in $\left[-\frac{\lambda_1+\beta_1}{2}, -\lambda_1\right] \cup \left[\lambda_1 \frac{\lambda_1+\beta_1}{2}\right]$. Since (2.3) implies that $\tilde{a}'$ is bounded far from x_0, by [**2**, Theorem 3.1] and [**2**, Remark 7] we get

$$\begin{aligned} &\int_0^T \int_{-\beta_1}^{\beta_1} \left[s\Theta (Z_x)^2 e^{2s\Phi} + s^3 \Theta^3 Z^2 e^{2s\Phi} \right] dxdt \le C \int_0^T \int_{-\beta_1}^{\beta_1} e^{2s\Phi} \bar{h}^2 dxdt \\ &\le C \int_0^T \int_{-\frac{\lambda_1+\beta_1}{2}}^{-\lambda_1} e^{2s\Phi}(W^2 + (W_x)^2) dxdt + C \int_0^T \int_{\lambda_1}^{\frac{\lambda_1+\beta_1}{2}} e^{2s\Phi}(W^2 + (W_x)^2) dxdt \\ &\le C \int_0^T \int_{-\beta}^{-\alpha} W^2 dxdt + C \int_0^T \int_\omega W^2 dxdt \le C \int_0^T \int_\omega v^2 dxdt, \end{aligned} \tag{2.7}$$

for some positive constants C, where Φ is the weight defined in (3.3) of [**2**] for the interval $[-\beta_1, \beta_1]$. Now, one can prove as in [**2**] that there exists a positive constant k, for example

$$k = \max\left\{ \max_{[0,\lambda_1]} a, \frac{(x_0)^2}{a(0)} \right\},$$

such that

$$a(x) e^{2s\varphi(t,x)} \le k e^{2s\Phi(t,x)} \quad \text{and} \quad \frac{(x-x_0)^2}{a(x)} e^{2s\varphi(t,x)} \le k e^{2s\Phi(t,x)}$$

for every $(t,x) \in [0,T] \times [0,\lambda_1]$. Thus, by (2.7), one has

$$\begin{aligned}
&\int_0^T \int_0^{\lambda_1} \Big(s\Theta a(v_x)^2 + s^3\Theta^3 \frac{(x-x_0)^2}{a} v^2\Big) e^{2s\varphi} dxdt \\
&= \int_0^T \int_0^{\lambda_1} \Big(s\Theta a(Z_x)^2 + s^3\Theta^3 \frac{(x-x_0)^2}{a} Z^2\Big) e^{2s\varphi} dxdt \\
&\le k \int_0^T \int_{-\beta_1}^{\beta_1} (s\Theta(Z_x)^2 e^{2s\Phi} dxdt + s^3\Theta^3 Z^2 e^{2s\Phi}) dxdt \le kC \int_0^T \int_\omega v^2 dxdt.
\end{aligned}$$

Hence, Lemma 5.1 follows. □

As said, Lemma 5.2 holds assuming (2.2) in place of (5.2).

PROOF OF LEMMA 5.2. Fix $\bar\lambda_i, \bar\beta_i \in (\lambda_i, \beta_i)$, $i = 1, 2$, with $\bar\lambda_i < \bar\beta_i$ and consider a smooth function $\xi : [0,1] \to [0,1]$ such that

$$\xi(x) = \begin{cases} 0 & x \in [0, \bar\lambda_1], \\ 1 & x \in [\tilde\lambda_1, \tilde\lambda_2], \\ 0 & x \in [\bar\beta_2, 1], \end{cases} \tag{2.8}$$

where $\tilde\lambda_i = (\bar\lambda_i + \bar\beta_i)/2$, $i = 1, 2$. Then, define $w := \xi v$, where v is any fixed solution of (5.7), so that w satisfies

$$\begin{cases} w_t + (aw_x)_x = f, & (t,x) \in Q_T, \\ w(t,0) = w(t,1) = 0, & t \in (0,T), \end{cases}$$

with $f^2 = ((a\xi_x v)_x + a\xi_x v_x)^2 \le C(v^2 + (v_x)^2)\chi_{\hat\omega}$, where $\hat\omega = (\bar\lambda_1, \tilde\lambda_1) \cup (\tilde\lambda_2, \bar\beta_2)$. By applying [**2**, Theorem 4.1], we have

$$\int_0^T \int_0^1 \Big(s\Theta a(w_x)^2 + s^3\Theta^3 \frac{(x-x_0)^2}{a} w^2\Big) e^{2s\varphi} \, dxdt \le C \int_0^T \int_0^1 e^{2s\varphi} f^2 dxdt, \tag{2.9}$$

for all $s \ge s_0$. Hence, by [**2**, Proposition 5.2] we find

$$\begin{aligned}
&\int_0^T \int_{\tilde\lambda_1}^{\tilde\lambda_2} \left(s\Theta a(v_x)^2 + s^3\Theta^3 \frac{(x-x_0)^2}{a} v^2\right) e^{2s\varphi} dxdt \\
&\le \int_0^T \int_0^1 \Big(s\Theta a(w_x)^2 + s^3\Theta^3 \frac{(x-x_0)^2}{a} w^2\Big) e^{2s\varphi} \, dxdt \\
&\le C \int_0^T \int_{\hat\omega} e^{2s\varphi}(v^2 + (v_x)^2) dxdt \le C \int_0^T \int_\omega v^2 dxdt.
\end{aligned} \tag{2.10}$$

Analogously, also for future purposes, we define a smooth function $\tau : [0,2] \to [0,1]$ such that

$$\tau(x) = \begin{cases} 0 & x \in [0, \bar\lambda_2] \cup [2 - \bar\lambda_2, 2], \\ 1 & x \in [\tilde\lambda_2, 2 - \tilde\lambda_2]. \end{cases} \tag{2.11}$$

Defining $z := \tau v$, then z satisfies

$$\begin{cases} z_t + (az_x)_x = h, & (t,x) \in (0,T) \times (\lambda_2, 1), \\ z(t,\lambda_2) = z(t,1) = 0, & t \in (0,T), \end{cases}$$

with $h := (a\tau_x v)_x + a\tau_x v_x$. By applying [**2**, Theorem 3.1] with $A = \lambda_2$, $B = 1$ and related weight Φ, we have

$$\int_0^T \int_{\lambda_2}^1 \left[s\Theta(z_x)^2 e^{2s\Phi} + s^3\Theta^3 z^2 e^{2s\Phi}\right] dxdt \le c \int_0^T \int_{\lambda_2}^1 e^{2s\Phi} h^2 dxdt, \tag{2.12}$$

for $s \ge s_0$ and $c > 0$. Notice that the boundary term is nonpositive and thus is neglected. Now, proceeding as in [**2**], there exists a constant $k > 0$ such that, by [**2**, Proposition 5.2], as above, we get

$$\begin{aligned}&\int_0^T \int_{\lambda_2}^1 \left(s\Theta a(z_x)^2 + s^3\Theta^3 \frac{(x-x_0)^2}{a} z^2\right) e^{2s\varphi} dxdt \\ &\le k \int_0^T \int_{\lambda_2}^1 \left[s\Theta(z_x)^2 + s^3\Theta^3 z^2\right] e^{2s\Phi} dxdt \le C \int_0^T \int_\omega v^2 dxdt,\end{aligned}$$

for a positive constant C and s large enough. Hence, we get

$$\begin{aligned}&\int_0^T \int_{\tilde\lambda_2}^1 \left(s\Theta a(v_x)^2 + s^3\Theta^3 \frac{(x-x_0)^2}{a} v^2\right) e^{2s\varphi} dxdt \\ &\le \int_0^T \int_{\lambda_2}^1 \left(s\Theta a(z_x)^2 + s^3\Theta^3 \frac{(x-x_0)^2}{a} z^2\right) e^{2s\varphi} dxdt \le C \int_0^T \int_\omega v^2 dxdt.\end{aligned} \tag{2.13}$$

Finally, consider a smooth function $\rho : [-1,1] \to [0,1]$ such that

$$\rho(x) = \begin{cases} 0 & x \in [-1, -\bar\beta_1], \\ 1 & x \in [-\tilde\lambda_1, \tilde\lambda_1], \\ 0 & x \in [\bar\beta_1, 1], \end{cases} \tag{2.14}$$

and define $Z := \rho W$, where W is defined in (2.5); thus Z satisfies (2.6) with $\bar h = (\tilde a \rho_x W)_x + \tilde a \rho_x W_x$. By applying [**2**, Theorem 3.1] with $A = -\beta_1$ and $B = \beta_1$ with the corresponding Φ, by [**2**, Propositions 5.2] we get

$$\begin{aligned}&\int_0^T \int_{-\beta_1}^{\beta_1} \left(s\Theta(Z_x)^2 + s^3\Theta^3 Z^2\right) e^{2s\Phi} dxdt \le c \int_0^T \int_{-\beta_1}^{\beta_1} e^{2s\Phi} \bar h^2 dxdt \\ &\le C \int_0^T \int_{-\bar\beta_1}^{-\tilde\lambda_1} (W^2 + (W_x)^2) e^{2s\Phi} dxdt + C \int_0^T \int_{\tilde\lambda_1}^{\bar\beta_1} (W^2 + (W_x)^2) e^{2s\Phi} dxdt \\ &\le C \int_0^T \int_{-\beta_1}^{-\lambda_1} W^2 dxdt + C \int_0^T \int_{\lambda_1}^{\beta_1} W^2 dxdt \le C \int_0^T \int_\omega v^2 dxdt,\end{aligned}$$

for some positive constants c, C and s large enough. Thus, as before, we get

$$\begin{aligned}&\int_0^T \int_0^{\tilde\lambda_1} \left(s\Theta a(v_x)^2 + s^3\Theta^3 \frac{(x-x_0)^2}{a} v^2\right) e^{2s\varphi} dxdt \\ &\le k \int_0^T \int_0^{\tilde\lambda_1} \left[s\Theta(Z_x)^2 e^{2s\Phi} + s^3\Theta^3 Z^2 e^{2s\Phi}\right] dxdt \\ &\le k \int_0^T \int_{-\beta_1}^{\beta_1} \left(s\Theta(Z_x)^2 + s^3\Theta^3 Z^2\right) e^{2s\Phi} dxdt \le C \int_0^T \int_\omega v^2 dxdt.\end{aligned} \tag{2.15}$$

The conclusion follows by (2.10), (2.13) and (2.15). $\square$

The rest of the proof of Proposition 5.1 is as in [**2**].

The non divergence case

Proposition 5.3 (and consequent Theorem 5.3) holds with assumption (2.2) in place of (5.2). In particular, Lemma 5.4 holds with (2.2) in place of (5.2).

PROOF OF LEMMA 5.4. Let $\tilde{\lambda}_i, \bar{\lambda}_i, \bar{\beta}_i \in (\lambda_i, \beta_i)$, $i = 1, 2$, be as in the proof of Lemma 5.2; consider ξ, τ, ρ, $\hat{\omega}$ defined therein and take $w = \xi v$, $z = \tau v$ and $Z = \rho W$, where v is any fixed solution of (5.38) and W is defined in (2.5). Hence, w, z and Z satisfy

$$\begin{cases} w_t + a w_{xx} = a(\xi_{xx} v + 2\xi_x v_x) =: f, & (t,x) \in (0,T) \times (0,1), \\ w(t,0) = w(t,1) = 0, & t \in (0,T), \end{cases} \tag{2.16}$$

$$\begin{cases} z_t + a z_{xx} = a(\tau_{xx} v + 2\tau_x v_x) =: h, & (t,x) \in (0,T) \times (\lambda_2,1), \\ z(t,\lambda_2) = z(t,1) = 0, & t \in (0,T), \end{cases} \tag{2.17}$$

and

$$\begin{cases} Z_t + \tilde{a} Z_{xx} = \tilde{a}\rho_{xx} W + 2\tilde{a}\rho_x W_x = \bar{h}, & (t,x) \in (0,T) \times (-\beta_1,\beta_1), \\ Z(t,-\beta_1) = Z(t,\beta_1) = 0, & t \in (0,T). \end{cases} \tag{2.18}$$

By applying [**2**, Theorem 4.2] to (2.16), by [**2**, Proposition 5.4], we have for all $s \geq s_0$

$$\begin{aligned} &\int_0^T \int_{\tilde{\lambda}_1}^{\tilde{\lambda}_2} \left(s\Theta (v_x)^2 + s^3 \Theta^3 \left(\frac{x - x_0}{a} \right)^2 v^2 \right) e^{2s\varphi} dxdt \\ &\leq \int_{Q_T} \left(s\Theta (w_x)^2 + s^3 \Theta^3 \left(\frac{x - x_0}{a} \right)^2 w^2 \right) e^{2s\varphi} \, dxdt \leq C \int_{Q_T} \frac{e^{2s\varphi}}{a} f^2 dxdt \\ &\leq C \int_0^T \int_{\hat{\omega}} e^{2s\varphi}(v^2 + (v_x)^2) dxdt \leq C \int_0^T \int_{\omega} \frac{v^2}{a} dxdt. \end{aligned} \tag{2.19}$$

Moreover, applying [**2**, Theorem 3.2] to (2.17) with $A = \lambda_2$, $B = 1$ and related Φ, we obtain again (2.12) with $h = a(\tau_{xx} v + 2\tau_x v_x)$.

Now, working with z, recalling what the support of τ is, we similarly get

$$\int_0^T \int_{\lambda_2}^1 \left(s\Theta (z_x)^2 + s^3 \Theta^3 z^2 \right) e^{2s\Phi} dxdt \leq C \int_0^T \int_{\omega} \frac{v^2}{a} dxdt. \tag{2.20}$$

As usual, we can find $k > 0$ such that for a positive constant C and s large enough

$$\begin{aligned} &\int_0^T \int_{\tilde{\lambda}_2}^1 \left(s\Theta (v_x)^2 + s^3 \Theta^3 \left(\frac{x - x_0}{a} \right)^2 v^2 \right) e^{2s\varphi} dxdt \\ &\leq k \int_0^T \int_{\lambda_2}^1 \left(s\Theta (z_x)^2 + s^3 \Theta^3 z^2 \right) e^{2s\Phi} dxdt \leq C \int_0^T \int_{\omega} \frac{v^2}{a} dxdt. \end{aligned} \tag{2.21}$$

Finally, applying [**2**, Theorem 3.2] to (2.18) with $A = -\beta_1$, $B = \beta_1$ and the associated Φ, we similarly get

(2.22)

$$
\begin{aligned}
&\int_0^T \int_0^{\tilde{\lambda}_1} \left(s\Theta a (v_x)^2 + s^3\Theta^3 \left(\frac{x - x_0}{a}\right)^2 v^2 \right) e^{2s\varphi} dxdt \\
&\le k \int_0^T \int_{-\beta_1}^{\beta_1} \left(s\Theta (Z_x)^2 + s^3\Theta^3 Z^2 \right) e^{2s\Phi} dxdt \le c \int_0^T \int_{-\beta_1}^{\beta_1} e^{2s\Phi} \bar{h}^2 dxdt \\
&\le C \int_0^T \int_{-\bar{\beta}_1}^{-\tilde{\lambda}_1} (W^2 + (W_x)^2) e^{2s\Phi} dxdt + C \int_0^T \int_{\tilde{\lambda}_1}^{\bar{\beta}_1} (W^2 + (W_x)^2) e^{2s\Phi} dxdt \\
&\text{(by [2, Propositions 5.4])} \\
&\le C \int_0^T \int_{-\beta_1}^{-\lambda_1} W^2 dxdt + C \int_0^T \int_{\lambda_1}^{\beta_1} W^2 dxdt \le C \int_0^T \int_\omega v^2 dxdt,
\end{aligned}
$$

for a positive constant C and s large enough. By (2.19), (2.21) and (2.22), the conclusion follows. □

The rest of the proof is the same as in [**2**].

CHAPTER 3

Correction to [1]

The new hypotheses for [**2**] imply changes in [**1**] for Proposition 4.1 and Theorem 4.1 only if Neumann boundary conditions - shortly (Nbc) - hold, and in Lemma 5.1, while Proposition 4.1 and Theorem 4.1 are correct for Dirichlet boundary conditions - shortly (Dbc). Hypothesis 4.1 must be changed in the (Nbc) case only in the part (a_1), but for completeness we give the entire

Hypothesis 4.1 - (Nbc). (a_1) $a \in W^{1,1}(0,1)$, there exist $\sigma \in (0,1)$, $\mathfrak{g} \in L^1(-\sigma, 1+\sigma)$, $\mathfrak{h} \in W^{1,\infty}(-\sigma, 1+\sigma)$ and two strictly positive constants $\mathfrak{g}_0$, $\mathfrak{h}_0$ such that $\mathfrak{g}(x) \geq \mathfrak{g}_0$ and

$$\frac{\tilde{a}'(x)}{2\sqrt{\tilde{a}(x)}}\left(\int_x^{1+\sigma} \mathfrak{g}(t)dt + \mathfrak{h}_0\right) + \sqrt{\tilde{a}(x)}\mathfrak{g}(x) = \mathfrak{h}(x) \quad \text{for a.e. } x \in [-\sigma, 1+\sigma], \tag{3.1}$$

where

$$\tilde{a}(x) = \begin{cases} a(-x), & x \in [-1,0], \\ a(x), & x \in [0,1], \\ a(2-x), & x \in [1,2], \end{cases} \tag{3.2}$$

or

(a_2) $a \in W^{1,\infty}(0,1)$.

Set $\tilde{\Phi}(t,x) := \Theta(t)\rho_{-\sigma,1+\sigma}(x)$ and $\Phi(t,x) := \Theta(t)\rho_{0,1}(x)$, where, for $A < B$,

$$\rho_{A,B}(x) := \begin{cases} -r\left[\displaystyle\int_A^x \frac{1}{\sqrt{\tilde{a}(t)}}\int_t^B \mathfrak{g}(s)ds dt + \int_A^x \frac{\mathfrak{h}_0}{\sqrt{\tilde{a}(t)}}dt\right] - \mathfrak{c}, & \text{in the (WD)}, \\ e^{r\zeta(x)} - \mathfrak{c}, & \text{in the (SD)}, \end{cases} \tag{3.3}$$

$$\zeta(x) = \mathfrak{d}\int_x^B \frac{1}{\tilde{a}(t)}dt,$$

with $\mathfrak{d} = \|\tilde{a}'\|_{L^\infty(A,B)}$, $r > 0$ and $\mathfrak{c} > 0$ is such that $\max\limits_{[A,B]} \rho_{A,B} < 0$.

REMARK 2. (1) $\rho_{-\sigma,1+\sigma}(x) \leq \rho_{0,1}(x)$ for all $x \in [0,1]$, thus in (3.3) it is sufficient to choose $\mathfrak{c}$ such that $\max\limits_{[0,1]} \rho_{0,1} < 0$ in order to ensure that $\rho_{-\sigma,1+\sigma} < 0$. Hence, $\tilde{\Phi}(t,x) \leq \Phi(t,x)$ for all $x \in [0,1]$,

(2) $\rho_{-\sigma,1+\sigma}(2-x) \leq \rho_{-\sigma,1+\sigma}(x) \leq \rho_{-\sigma,1+\sigma}(-x)$ for all $x \in [0,1]$. Hence,

$$\tilde{\Phi}(t, 2-x) \leq \tilde{\Phi}(t,x) \leq \tilde{\Phi}(t,-x) \quad \text{for all } x \in [0,1].$$

Similar estimates hold for $\rho_{A,B}$ for any A, B.

Proposition 4.1 (Nondegenerate Carleman estimate - (Nbc)) *Assume Hypothesis* 4.1 *- (Nbc) case. Let z solve* (4.4). *Then, for any $\delta \in (0,\sigma)$ with $\delta+\sigma<1$, there exist three positive constants $C = C(\sigma,\delta)$, r and s_0 such that for any $s \geq s_0$*
(3.4)

$$\int_{Q_T}\left(s\Theta(z_x)^2+s^3\Theta^3z^2\right)e^{2s\tilde{\Phi}(t,x)}dxdt \leq C\left(\int_0^T\int_0^{\sigma+\delta}z^2dxdt+C\int_0^T\int_{1-\sigma-\delta}^1 z^2dxdt\right.$$

$$\left.+\int_{Q_T}h^2e^{2s\Phi(t,x)}dxdt+\int_0^T\int_0^{\sigma}h^2e^{2s\tilde{\Phi}(t,-x)}dxdt\right).$$

Proof. Take W as in (4.10), fix δ as indicated and consider a smooth function $\xi:[-1,2]\to\mathbb{R}$ such that $\xi\equiv 1$ in $[-\delta,1+\delta]$ and $\xi\equiv 0$ in $[-1,-\sigma]\cup[1+\sigma,2]$. Set $Z:=\xi W$, so that Z solves

$$\begin{cases} Z_t+\tilde{a}Z_{xx}+\lambda\dfrac{Z}{\tilde{b}}=H, & (t,x)\in(0,T)\times(-\sigma,1+\sigma),\\ Z(t,-\sigma)=Z(t,1+\sigma)=0, & t\in(0,T),\end{cases}$$

with $H:=\xi\tilde{h}+\tilde{a}(\xi_{xx}W+2\xi_xW_x)$, $\tilde{b}$ and $\tilde{h}$ defined in (4.12) and (4.13), respectively. Thus, we can apply [**1**, Proposition 4.1] in the (Dbc) case with $(-\sigma,1+\sigma)$ in place of $(0,1)$, $\tilde{a}$ in place of a and with $Z_x(t,-\sigma)=Z_x(t,1+\sigma)=0$. Thus, [**1**, Hypothesis 4.1] holds in the (Dbc) case. Hence, we find two positive constants C (depending on σ) and s_0 (s_0 sufficiently large), such that Z satisfies, for all $s\geq s_0$

$$\int_0^T\int_{-\sigma}^{1+\sigma}\left(s\Theta(Z_x)^2+s^3\Theta^3Z^2\right)e^{2s\tilde{\Phi}}dxdt \leq C\int_0^T\int_{-\sigma}^{1+\sigma}H^2e^{2s\tilde{\Phi}}dxdt$$

$$\leq C\int_0^T\int_{-\sigma}^{1+\sigma}\tilde{h}^2e^{2s\tilde{\Phi}}dxdt$$

$$+C\left(\int_0^T\int_{-\sigma}^{-\delta}(W^2+W_x^2)e^{2s\tilde{\Phi}}dxdt+\int_0^T\int_{1+\delta}^{1+\sigma}(W^2+W_x^2)e^{2s\tilde{\Phi}}dxdt\right) \tag{3.5}$$

(by [**1**, Remark 7])

$$\leq C\int_0^T\int_{-\sigma}^{1+\sigma}\tilde{h}^2e^{2s\tilde{\Phi}}dxdt+C\int_0^T\int_0^{\sigma+\delta}z^2dxdt+C\int_0^T\int_{1-\sigma-\delta}^1z^2dxdt.$$

Now, observe that

$$\int_0^T\int_{-\sigma}^{1+\sigma}\tilde{h}^2(t,x)e^{2s\tilde{\Phi}(t,x)}dxdt \tag{3.6}$$

$$\leq\int_0^T\int_0^{\sigma}h^2(t,x)e^{2s\tilde{\Phi}(t,-x)}dxdt+\int_{Q_T}h^2(t,x)e^{2s\Phi(t,x)}dxdt.$$

Indeed, using a change of variable, the definition of $\tilde h$ and Remark 2, we get

$$\int_{-\sigma}^{1+\sigma}\tilde h^2(t,x)e^{2s\tilde\Phi(t,x)}dx=\int_{-\sigma}^{0}\tilde h^2(t,x)e^{2s\tilde\Phi(t,x)}dx$$
$$+\int_0^1\tilde h^2(t,x)e^{2s\tilde\Phi(t,x)}dx+\int_1^{1+\sigma}\tilde h^2(t,x)e^{2s\tilde\Phi(t,x)}dx$$
$$=\int_0^{\sigma}\tilde h^2(t,-y)e^{2s\tilde\Phi(t,-y)}dy+\int_0^1 h^2(t,x)e^{2s\tilde\Phi(t,x)}dx+\int_{1-\sigma}^{1}\tilde h^2(t,2-y)e^{2s\tilde\Phi(t,2-y)}dy$$
$$\le\int_0^{\sigma}h^2(t,y)e^{2s\tilde\Phi(t,-y)}dy+\int_0^1 h^2(t,x)e^{2s\Phi(t,x)}dx+\int_{1-\sigma}^{1}h^2(t,y)e^{2s\Phi(t,y)}dy.$$

Thus, by (3.5) and (3.6), we obtain

(3.7)
$$\int_0^T\int_{-\sigma}^{1+\sigma}\left(s\Theta(Z_x)^2+s^3\Theta^3Z^2\right)e^{2s\tilde\Phi}dxdt\le C\left(\int_0^T\int_0^{\sigma+\delta}z^2dxdt+\int_0^T\int_{1-\sigma-\delta}^{1}z^2dxdt\right)$$
$$+C\int_0^T\int_0^{\sigma}h^2(t,x)e^{2s\tilde\Phi(t,-x)}dxdt+C\int_{Q_T}h^2(t,x)e^{2s\Phi(t,x)}dxdt.$$

As a consequence,

$$\int_{Q_T}\left(s\Theta(z_x)^2+s^3\Theta^3z^2\right)e^{2s\tilde\Phi}dxdt\le\int_0^T\int_{-\sigma}^{1+\sigma}\left(s\Theta(Z_x)^2+s^3\Theta^3Z^2\right)e^{2s\tilde\Phi}dxdt$$
$$\le C\left(\int_0^T\int_0^{\sigma+\delta}z^2dxdt+\int_0^T\int_{1-\sigma-\delta}^{1}z^2dxdt\right.$$
$$\left.+\int_0^T\int_0^{\sigma}h^2(t,x)e^{2s\tilde\Phi(t,-x)}dxdt+\int_{Q_T}h^2(t,x)e^{2s\Phi(t,x)}dxdt\right),$$

for a positive constant C depending on σ and δ. The claim follows. □

REMARK 3. Proposition 4.1 still holds if we replace $[0,1]$ with a general interval $[A,B]$ where a and b satisfy Hypothesis 4.1 in $[A-\sigma,B+\sigma]$. In this case $\int_0^T\int_0^\sigma h^2(t,x)e^{2s\tilde\Phi(t,-x)}dxdt$ is replaced by $\int_0^T\int_A^{A+\sigma}h^2(t,x)e^{2s\tilde\Phi(t,2A-x)}dxdt$ with $\tilde\Phi:=\Theta(t)\rho_{A-\sigma,B+\sigma}$.

Again, in the weak degenerate case, there is a change only in the (Nbc) case:

Hypothesis 4.2 - (Nbc). The assumption is as in [**1**] and in addition: if Hypothesis 1.1 or Hypothesis 1.3 of [**1**] holds, there exist $B_1\in(0,x_0)$, $B_2\in(1,2-x_0)$, two functions $\mathfrak{g}\in L^\infty_{\rm loc}((-x_0,2-x_0)\setminus\{x_0\})$, $\mathfrak{h}(\cdot,B_i)\in W^{1,\infty}_{\rm loc}((-x_0,2-x_0)\setminus\{x_0\})$ and two strictly positive constants $\mathfrak{g}_0$, $\mathfrak{h}_0$ such that $\mathfrak{g}(x)\ge\mathfrak{g}_0$ and

$$\frac{\tilde a'(x)}{2\sqrt{\tilde a(x)}}\left(\int_x^{B_i}\mathfrak{g}(t)dt+\mathfrak{h}_0\right)+\sqrt{\tilde a(x)}\mathfrak{g}(x)=\mathfrak{h}(x,B_i)\tag{3.8}$$

with $i=1,2$, $-x_0<x<B_1$ or $x_0<x<B_2$, and $\tilde a$ defined as in (3.2).

Using $\varphi:=\Theta\psi$ and the space $\mathcal{V}$ introduced in [**1**], Theorem 4.1 in the (Nbc) is improved in this way:

Theorem 4.1 - (Nbc). *Assume Hypothesis* 4.2 - *(Nbc)* . *Let* $\omega\subset\subset(0,1)$ *be an open interval containing* x_0, B_1 *and* $2-B_2$, *or let* $\omega=\omega_1\cup\omega_2$, *where* $\omega_i=(\lambda_i,\beta_i)\subset$

$(0,1)$, $i=1,2$, $\beta_1 \le B_1$ and $2-B_2 \le \lambda_2$. Then, there exist two positive constants C and s_0 (depending on λ) such that every solution $v \in \mathcal{V}$ of (4.26) satisfies, for all $s \ge s_0$,

(3.9)
$$\begin{aligned}
&\int_{Q_T}\left(s\Theta(v_x)^2 + s^3\Theta^3\left(\frac{x-x_0}{a}\right)^2 v^2\right)e^{2s\varphi}dxdt\\
&\le C\left(\int_0^T\int_\omega v^2dxdt + \int_{Q_T}\frac{h^2(t,x)}{a}e^{2s\varphi}dxdt\right)\\
&+C\left(\int_0^T\int_0^{B_1}\frac{h^2(t,x)}{a}e^{2s\Phi_1(t,-x)}dxdt + \int_0^T\int_{2-B_2}^1\frac{h^2(t,x)}{a}e^{2s\Phi_2(t,x)}dxdt\right),
\end{aligned}$$

where $\Phi_1(t,x) := \Theta(t)\rho_{-B_1,B_1}(x)$ and $\Phi_2(t,x) := \Theta(t)\rho_{2-B_2,B_2}(x)$.

PROOF. First, assume that $x_0, B_1, 2-B_2 \in \omega$. Then, we can fix two subintervals $\omega_1 = (\lambda_1,\beta_1) \subset\subset (0,x_0), \omega_2 = (\lambda_2,\beta_2) \subset\subset (x_0,1)$, with $\beta_1 = B_1$ and $\lambda_2 = 2-B_2$. Let v solve (4.26), take ξ as in (2.8) and define $w := \xi v$, so that w satisfies

$$\begin{cases} w_t + a(x)w_{xx} + \dfrac{\lambda}{b(x)}w = \bar{h}, & (t,x) \in Q_T,\\ w(t,0) = w(t,1) = 0, & t \in (0,T), \end{cases} \tag{3.10}$$

where $\bar{h} := \xi h + a(\xi_{xx}v + 2\xi_x v_x)$. Using [**1**, Theorem 4.1] in the (Dbc) case, one has

(3.11)
$$\int_0^T\int_0^1\left(s\Theta a(w_x)^2 + s^3\Theta^3\left(\frac{x-x_0}{a}\right)^2 w^2\right)e^{2s\varphi}\,dxdt \le C\int_0^T\int_0^1\frac{\bar{h}^2}{a}e^{2s\varphi}dxdt,$$

for all $s \ge s_0$. Hence, by the Caccioppoli inequality [**1**, Proposition 5.2], proceeding as in (2.10), we find

$$\begin{aligned}
&\int_0^T\int_{\tilde{\lambda}_1}^{\tilde{\lambda}_2}\left(s\Theta a(v_x)^2 + s^3\Theta^3\left(\frac{x-x_0}{a}\right)^2 v^2\right)e^{2s\varphi}dxdt\\
&\le C\int_0^T\int_0^1\frac{h^2}{a}e^{2s\varphi}dxdt + C\int_0^T\int_\omega v^2dxdt.
\end{aligned} \tag{3.12}$$

Defining $z := \tau W$, W being defined in (4.25) and τ in (2.11), then z satisfies

$$\begin{cases} z_t + \tilde{a}(x)z_{xx} + \dfrac{\lambda}{\tilde{b}(x)}z = f, & (t,x) \in (0,T)\times(\lambda_2, 2-\lambda_2),\\ z(t,\lambda_2) = z(t,2-\lambda_2) = 0, & t \in (0,T), \end{cases} \tag{3.13}$$

with $f := \tau\tilde{h} + \tilde{a}(\tau_{xx}W + 2\tau_x W_x)$. Then, [**1**, Proposition 4.1] in the (Dbc) case can be applied; by Remark 2 and the Caccioppoli inequality in the non degenerate case

[**1**, Remark 7], we get

(3.14)

$$\int_0^T\int_{\lambda_2}^{2-\lambda_2}\left[s\Theta(z_x)^2e^{2s\Phi_2}+s^3\Theta^3z^2e^{2s\Phi_2}\right]dxdt\le C\int_0^T\int_{\lambda_2}^{2-\lambda_2}f^2(t,x)e^{2s\Phi_2(t,x)}dxdt$$
$$\le C\int_0^T\int_{\lambda_2}^1\frac{h^2(t,x)}{a}e^{2s\Phi_2(t,x)}dxdt+C\int_0^T\int_{\tilde\omega}(W^2+W_x^2)e^{2s\Phi_2(t,x)}dxdt$$
$$\le C\int_0^T\int_{\lambda_2}^1\frac{h^2(t,x)}{a}e^{2s\Phi_2(t,x)}dxdt+C\int_0^T\int_{\omega}v^2dxdt,$$

where $\tilde\omega=(\bar\lambda_2,\tilde\lambda_2)\cup(2-\tilde\lambda_2,2-\bar\lambda_2)$. Proceeding as before, there exists $k>0$ such that, for a positive constant C and s large enough,

$$\int_0^T\int_{\lambda_2}^1\left(s\Theta(z_x)^2+s^3\Theta^3\left(\frac{x-x_0}{a}\right)^2z^2\right)e^{2s\varphi}dxdt$$
$$\le k\int_0^T\int_{\lambda_2}^{2-\lambda_2}\left[s\Theta(z_x)^2e^{2s\Phi_2}+s^3\Theta^3z^2e^{2s\Phi_2}\right]dxdt$$
$$\le C\int_0^T\int_{\lambda_2}^1\frac{h^2(t,x)}{a}e^{2s\Phi_2(t,x)}dxdt+C\int_0^T\int_{\omega}v^2dxdt.$$

Hence, by definition of z and by the inequality above, we get

$$\int_0^T\int_{\tilde\lambda_2}^1\left(s\Theta(v_x)^2+s^3\Theta^3\left(\frac{x-x_0}{a}\right)^2v^2\right)e^{2s\varphi}dxdt$$
$$=\int_0^T\int_{\tilde\lambda_2}^1\left(s\Theta(z_x)^2+s^3\Theta^3\left(\frac{x-x_0}{a}\right)^2z^2\right)e^{2s\varphi}dxdt \tag{3.15}$$
$$\le C\int_0^T\int_{\lambda_2}^1\frac{h^2(t,x)}{a}e^{2s\Phi_2(t,x)}dxdt+C\int_0^T\int_{\omega}v^2dxdt.$$

Finally, defining $Z:=\rho W$, where ρ is given in (2.14), then Z satisfies

$$\begin{cases}Z_t+\tilde aZ_{xx}+\lambda\dfrac{Z}{\tilde b}=\bar h, & (t,x)\in(0,T)\times(-\beta_1,\beta_1),\\ Z(t,-\beta_1)=Z(t,\beta_1)=0, & t\in(0,T),\end{cases}\tag{3.16}$$

with $\bar h:=\rho\tilde h+\tilde a(\rho_{xx}W+2\rho_xW_x)$. Proceeding as before, by [**1**, Proposition 4.1] in the (Dbc) case, Remark 2 and [**1**, Remark 7], one has

$$\int_0^T\int_0^{\tilde\lambda_1}\left(s\Theta(v_x)^2+s^3\Theta^3\left(\frac{x-x_0}{a}\right)^2v^2\right)e^{2s\varphi}dxdt$$
$$\le C\int_0^T\int_{-\beta_1}^{\beta_1}\left[s\Theta(Z_x)^2e^{2s\Phi_1}+s^3\Theta^3Z^2e^{2s\Phi_1}\right]dxdt \tag{3.17}$$
$$\le C\int_0^T\int_0^{\beta_1}\frac{h^2(t,x)}{a}e^{2s\Phi_1(t,-x)}dxdt+C\int_0^T\int_{\omega}v^2dxdt.$$

The conclusion follows by (2.10) and (3.15).

Nothing changes in the proof if $\omega=\omega_1\cup\omega_2$ and each of these intervals lye on different sides of x_0, as the assumption implies. □

0.1. Applications to observability inequality. In this subsection we shall make the necessary changes according to the new hypotheses. For this, we assume that the control set ω satisfies the following assumption:

Hypothesis 5.1. (2.1) or (2.2) hold.

Then, [**1**, Hypothesis 5.2] reads:

Hypothesis 5.2 Hypothesis 4.2 of [**1**] is satisfied. Moreover, under Hypothesis 1.1 or Hypothesis 1.3, condition (3.8) holds with $B_1 = \beta_1$ and $B_2 = 2 - \lambda_2$.

With this changes, Proposition 5.1 still holds true and the only changes are in the proof of Lemma 5.1.

Proof of Lemma 5.1 if (2.2) holds. As in the proof of Lemma 5.2 in Section 2 above, fix $\tilde{\lambda}_i, \bar{\lambda}_i, \bar{\beta}_i \in (\lambda_i, \beta_i)$, $i = 1, 2$, and smooth functions ξ, ρ, τ as given in (2.8), (2.14), (2.11). Set $w := \xi v$, so that w satisfies

$$\begin{cases} w_t + aw_{xx} + \lambda \dfrac{w}{b} = a(\xi_{xx} v + 2\xi_x v_x) =: f, & (t,x) \in Q_T, \\ w(t,0) = w(t,1) = 0, & t \in (0,T). \end{cases}$$

Applying [**1**, Theorem 4.1] with (Dbc), there exists two positive constants C and s_0 such that, for all $s \geq s_0$,

$$\int_{Q_T} \left(s\Theta (w_x)^2 + s^3 \Theta^3 \left(\frac{x - x_0}{a} \right)^2 w^2 \right) e^{2s\varphi}\, dxdt \leq C \int_{Q_T} \frac{e^{2s\varphi}}{a} f^2 dxdt. \tag{3.18}$$

Then, (2.19) holds.

Now, set $z := \tau W$, where W is defined in (2.5) or in (4.25) if (Dbc) or (Nbc) hold, respectively, so that z satisfies (3.13) with $\tilde{h} = 0$. Then (3.15) holds with $h = 0$.

Finally, set $Z := \rho W$, so that Z satisfies (3.16) with $\bar{h} = \tilde{a}(\rho_{xx} W + 2\rho_x W_x)$. By applying [**1**, Proposition 4.1] in $(-\beta_1, \beta_1)$ in the (Dbc) case, together with [**1**, Remark 7], we get

$$\begin{aligned} &\int_0^T \int_{-\beta_1}^{\beta_1} \left(s\Theta (Z_x)^2 + s^3 \Theta^3 Z^2 \right) e^{2s\Phi_1} dxdt \\ &\leq C \int_0^T \int_{-\bar{\beta}_1}^{-\tilde{\lambda}_1} (W^2 + (W_x)^2) e^{2s\Phi_1} dxdt + C \int_0^T \int_{\tilde{\lambda}_1}^{\bar{\beta}_1} (W^2 + (W_x)^2) e^{2s\Phi_1} dxdt \\ &\leq C \int_0^T \int_{-\beta_1}^{-\lambda_1} W^2 dxdt + C \int_0^T \int_{\lambda_1}^{\beta_1} W^2 dxdt \leq C \int_0^T \int_\omega v^2 dxdt \leq C \int_0^T \int_\omega \frac{v^2}{a} dxdt, \end{aligned}$$

for some positive constants c, C and s large enough. Hence, (3.17) holds with $h = 0$. Therefore, by (2.19) and (3.15), (3.17) with $h = 0$, the conclusion follows. □

Proof of Lemma 5.1 if (2.1) holds. Let ω_1, ω_2 be as in Remark 1. Then (5.27) holds. Now, take $z = \tau W$, with τ as in (2.11) and W as in the previous case; hence, z satisfies (3.13) with $f = \tilde{a}(\tau_{xx} W + 2\tau_x W_x)$, and so (3.15) holds with $h = 0$.

Finally, we consider $Z := \rho W$ with ρ as in (2.14). Then, Z satisfies (3.16) with $\bar{h} = \tilde{a}(\rho_{xx} W + 2\rho_x W_x)$. Thus, as above, (3.17) holds with $h = 0$. Hence, by (5.27) and (3.15), (3.17) with $h = 0$, the conclusion follows. □

CHAPTER 4

Correction to [3]

Here Hypothesis 4.1(ii) is replaced by (2.2), and so Lemma 4.7 and Proposition 4.4 still hold true.

Acknowledgments

The authors wish to thank Piermarco Cannarsa and Roberto Guglielmi for pointing out the mistake in [**2**].

Bibliography

[1] Genni Fragnelli, *Interior degenerate/singular parabolic equations in nondivergence form: well-posedness and Carleman estimates*, J. Differential Equations **260** (2016), no. 2, 1314–1371, DOI 10.1016/j.jde.2015.09.019. MR3419728

[2] Genni Fragnelli and Dimitri Mugnai, *Carleman estimates, observability inequalities and null controllability for interior degenerate nonsmooth parabolic equations*, Mem. Amer. Math. Soc. **242** (2016), no. 1146, v+84, DOI 10.1090/memo/1146. MR3498150

[3] Genni Fragnelli and Dimitri Mugnai, *Carleman estimates for singular parabolic equations with interior degeneracy and non-smooth coefficients*, Adv. Nonlinear Anal. **6** (2017), no. 1, 61–84, DOI 10.1515/anona-2015-0163. MR3604938

Editorial Information

To be published in the *Memoirs*, a paper must be correct, new, nontrivial, and significant. Further, it must be well written and of interest to a substantial number of mathematicians. Piecemeal results, such as an inconclusive step toward an unproved major theorem or a minor variation on a known result, are in general not acceptable for publication.

Papers appearing in *Memoirs* are generally at least 80 and not more than 200 published pages in length. Papers less than 80 or more than 200 published pages require the approval of the Managing Editor of the Transactions/Memoirs Editorial Board. Published pages are the same size as those generated in the style files provided for $\mathcal{A}\mathcal{M}\mathcal{S}$-LaTeX or $\mathcal{A}\mathcal{M}\mathcal{S}$-TeX.

Information on the backlog for this journal can be found on the AMS website starting from `http://www.ams.org/memo`.

A Consent to Publish is required before we can begin processing your paper. After a paper is accepted for publication, the Providence office will send a Consent to Publish and Copyright Agreement to all authors of the paper. By submitting a paper to the *Memoirs*, authors certify that the results have not been submitted to nor are they under consideration for publication by another journal, conference proceedings, or similar publication.

Information for Authors

Memoirs is an author-prepared publication. Once formatted for print and on-line publication, articles will be published as is with the addition of AMS-prepared frontmatter and backmatter. Articles are not copyedited; however, confirmation copy will be sent to the authors.

Initial submission. The AMS uses Centralized Manuscript Processing for initial submissions. Authors should submit a PDF file using the Initial Manuscript Submission form found at `www.ams.org/submission/memo`, or send one copy of the manuscript to the following address: Centralized Manuscript Processing, MEMOIRS OF THE AMS, 201 Charles Street, Providence, RI 02904-2294 USA. If a paper copy is being forwarded to the AMS, indicate that it is for *Memoirs* and include the name of the corresponding author, contact information such as email address or mailing address, and the name of an appropriate Editor to review the paper (see the list of Editors below).

The paper must contain a *descriptive title* and an *abstract* that summarizes the article in language suitable for workers in the general field (algebra, analysis, etc.). The *descriptive title* should be short, but informative; useless or vague phrases such as "some remarks about" or "concerning" should be avoided. The *abstract* should be at least one complete sentence, and at most 300 words. Included with the footnotes to the paper should be the 2010 *Mathematics Subject Classification* representing the primary and secondary subjects of the article. The classifications are accessible from `www.ams.org/msc/`. The Mathematics Subject Classification footnote may be followed by a list of *key words and phrases* describing the subject matter of the article and taken from it. Journal abbreviations used in bibliographies are listed in the latest *Mathematical Reviews* annual index. The series abbreviations are also accessible from `www.ams.org/msnhtml/serials.pdf`. To help in preparing and verifying references, the AMS offers MR Lookup, a Reference Tool for Linking, at `www.ams.org/mrlookup/`.

Electronically prepared manuscripts. The AMS encourages electronically prepared manuscripts, with a strong preference for $\mathcal{A}\mathcal{M}\mathcal{S}$-LaTeX. To this end, the Society has prepared $\mathcal{A}\mathcal{M}\mathcal{S}$-LaTeX author packages for each AMS publication. Author packages include instructions for preparing electronic manuscripts, samples, and a style file that generates the particular design specifications of that publication series. Though $\mathcal{A}\mathcal{M}\mathcal{S}$-LaTeX is the highly preferred format of TeX, author packages are also available in $\mathcal{A}\mathcal{M}\mathcal{S}$-TeX.

Authors may retrieve an author package for *Memoirs of the AMS* from `www.ams.org/journals/memo/memoauthorpac.html`. The *AMS Author Handbook* is available in PDF format from the author package link. The author package can also be obtained free of charge by sending email to tech-support@ams.org or from the Publication Division, American Mathematical Society, 201 Charles St., Providence, RI 02904-2294, USA. When requesting an author package, please specify AMS-LaTeX or AMS-TeX and the publication in which your paper will appear. Please be sure to include your complete mailing address.

After acceptance. The source files for the final version of the electronic manuscript should be sent to the Providence office immediately after the paper has been accepted for publication. The author should also submit a PDF of the final version of the paper to the editor, who will forward a copy to the Providence office.

Accepted electronically prepared files can be submitted via the web at www.ams.org/submit-book-journal/, sent via FTP, or sent on CD to the Electronic Prepress Department, American Mathematical Society, 201 Charles Street, Providence, RI 02904-2294 USA. TeX source files and graphic files can be transferred over the Internet by FTP to the Internet node ftp.ams.org (130.44.1.100). When sending a manuscript electronically via CD, please be sure to include a message indicating that the paper is for the *Memoirs*.

Electronic graphics. Comprehensive instructions on preparing graphics are available at www.ams.org/authors/journals.html. A few of the major requirements are given here.

Submit files for graphics as EPS (Encapsulated PostScript) files. This includes graphics originated via a graphics application as well as scanned photographs or other computer-generated images. If this is not possible, TIFF files are acceptable as long as they can be opened in Adobe Photoshop or Illustrator.

Authors using graphics packages for the creation of electronic art should also avoid the use of any lines thinner than 0.5 points in width. Many graphics packages allow the user to specify a "hairline" for a very thin line. Hairlines often look acceptable when proofed on a typical laser printer. However, when produced on a high-resolution laser imagesetter, hairlines become nearly invisible and will be lost entirely in the final printing process.

Screens should be set to values between 15% and 85%. Screens which fall outside of this range are too light or too dark to print correctly. Variations of screens within a graphic should be no less than 10%.

Any graphics created in color will be rendered in grayscale for the printed version unless color printing is authorized by the Managing Editor and the Publisher. In general, color graphics will appear in color in the online version.

Inquiries. Any inquiries concerning a paper that has been accepted for publication should be sent to memo-query@ams.org or directly to the Electronic Prepress Department, American Mathematical Society, 201 Charles St., Providence, RI 02904-2294 USA.

Editors

This journal is designed particularly for long research papers, normally at least 80 pages in length, and groups of cognate papers in pure and applied mathematics. Papers intended for publication in the *Memoirs* should be addressed to one of the following editors. The AMS uses Centralized Manuscript Processing for initial submissions to AMS journals. Authors should follow instructions listed on the Initial Submission page found at www.ams.org/memo/memosubmit.html.

Managing Editor: Henri Darmon, Department of Mathematics, McGill University, Montreal, Quebec H3A 0G4, Canada; e-mail: darmon@math.mcgill.ca

1. GEOMETRY, TOPOLOGY & LOGIC
   Coordinating Editor: Richard Canary, Department of Mathematics, University of Michigan, Ann Arbor, MI 48109-1043 USA; e-mail: canary@umich.edu

   Algebraic topology, Michael Hill, Department of Mathematics, University of California Los Angeles, Los Angeles, CA 90095 USA; e-mail: mikehill@math.ucla.edu

   Logic, Mariya Ivanova Soskova, Department of Mathematics, University of Wisconsin–Madison, Madison, WI 53706 USA; e-mail: msoskova@math.wisc.edu

   Low-dimensional topology and geometric structures, Richard Canary

   Symplectic geometry, Yael Karshon, School of Mathematical Sciences, Tel-Aviv University, Tel Aviv, Israel; and Department of Mathematics, University of Toronto, Toronto, Ontario M5S 2E4, Canada; e-mail: karshon@math.toronto.edu

2. ALGEBRA AND NUMBER THEORY
   Coordinating Editor: Henri Darmon, Department of Mathematics, McGill University, Montreal, Quebec H3A 0G4, Canada; e-mail: darmon@math.mcgill.ca

   Algebra, Radha Kessar, Department of Mathematics, City, University of London, London EC1V 0HB, United Kingdom; e-mail: radha.kessar.1@city.ac.uk

   Algebraic geometry, Lucia Caporaso, Department of Mathematics and Physics, Roma Tre University, Largo San Leonardo Murialdo, I-00146 Rome, Italy; e-mail: LCedit@mat.uniroma3.it

   Analytic number theory, Lillian B. Pierce, Department of Mathematics, Duke University, 120 Science Drive Box 90320, Durham, NC 27708 USA; e-mail: pierce@math.duke.edu

   Arithmetic geometry, Ted C. Chinburg, Department of Mathematics, University of Pennsylvania, Philadelphia, PA 19104-6395 USA; e-mail: ted@math.upenn.edu

   Commutative algebra, Irena Peeva, Department of Mathematics, Cornell University, Ithaca, NY 14853 USA; e-mail: irena@math.cornell.edu

   Number theory, Henri Darmon

3. GEOMETRIC ANALYSIS & PDE
   Coordinating Editor: Alexander A. Kiselev, Department of Mathematics, Duke University, 120 Science Drive, Rm 117 Physics Bldg, Durham, NC 27708 USA; e-mail: kiselev@math.duke.edu

   Differential geometry and geometric analysis, Ailana M. Fraser, Department of Mathematics, University of British Columbia, 1984 Mathematics Road, Room 121, Vancouver BC V6T 1Z2, Canada; e-mail: afraser@math.ubc.ca

   Harmonic analysis and partial differential equations, Monica Visan, Department of Mathematics, University of California Los Angeles, 520 Portola Plaza, Los Angeles, CA 90095 USA; e-mail: visan@math.ucla.edu

   Partial differential equations and functional analysis, Alexander A. Kiselev

   Real analysis and partial differential equations, Joachim Krieger, Bâtiment de Mathématiques, École Polytechnique Fédérale de Lausanne, Station 8, 1015 Lausanne Vaud, Switzerland; e-mail: joachim.krieger@epfl.ch

4. ERGODIC THEORY, DYNAMICAL SYSTEMS & COMBINATORICS
   Coordinating Editor: Vitaly Bergelson, Department of Mathematics, Ohio State University, 231 W. 18th Avenue, Columbus, OH 43210 USA; e-mail: vitaly@math.ohio-state.edu

   Algebraic and enumerative combinatorics, Jim Haglund, Department of Mathematics, University of Pennsylvania, Philadelphia, PA 19104 USA; e-mail: jhaglund@math.upenn.edu

   Probability theory, Robin Pemantle, Department of Mathematics, University of Pennsylvania, 209 S. 33rd Street, Philadelphia, PA 19104 USA; e-mail: pemantle@math.upenn.edu

   Dynamical systems and ergodic theory, Ian Melbourne, Mathematics Institute, University of Warwick, Coventry CV4 7AL, United Kingdom; e-mail: I.Melbourne@warwick.ac.uk

   Ergodic theory and combinatorics, Vitaly Bergelson

5. ANALYSIS, LIE THEORY & PROBABILITY
   Coordinating Editor: Stefaan Vaes, Department of Mathematics, Katholieke Universiteit Leuven, Celestijnenlaan 200B, B-3001 Leuven, Belgium; e-mail: stefaan.vaes@wis.kuleuven.be

   Functional analysis and operator algebras, Stefaan Vaes

   Harmonic analysis, PDEs, and geometric measure theory, Svitlana Mayboroda, School of Mathematics, University of Minnesota, 206 Church Street SE, 127 Vincent Hall, Minneapolis, MN 55455 USA; e-mail: svitlana@math.umn.edu

   Probability theory and stochastic analysis, Davar Khoshnevisan, Department of Mathematics, The University of Utah, Salt Lake City, UT 84112 USA; e-mail: davar@math.utah.edu

Selected Published Titles in This Series

1323 **Shu Kawaguchi and Kazuhiko Yamaki,** Effective Faithful Tropicalizations Associated to Linear Systems on Curves, 2021

1322 **D. Bulacu and B. Torrecillas,** Galois and Cleft Monoidal Cowreaths. Applications, 2021

1321 **Christian Haase, Andreas Paffenholz, Lindsey C. Piechnik, and Francisco Santos,** Existence of Unimodular Triangulations–Positive Results, 2021

1320 **Th. Heidersdorf and R. Weissauer,** Cohomological Tensor Functors on Representations of the General Linear Supergroup, 2021

1319 **Abed Bounemoura and Jacques Féjoz,** Hamiltonian Perturbation Theory for Ultra-Differentiable Functions, 2021

1318 **Chao Wang, Zhifei Zhang, Weiren Zhao, and Yunrui Zheng,** Local Well-Posedness and Break-Down Criterion of the Incompressible Euler Equations with Free Boundary, 2021

1317 **Eric M. Rains and S. Ole Warnaar,** Bounded Littlewood Identities, 2021

1316 **Ulrich Bunke and David Gepner,** Differential Function Spectra, the Differential Becker-Gottlieb Transfer, and Applications to Differential Algebraic K-Theory, 2021

1315 **S. Grivaux, É. Matheron, and Q. Menet,** Linear Dynamical Systems on Hilbert Spaces: Typical Properties and Explicit Examples, 2021

1314 **Pierre Albin, Frédéric Rochon, and David Sher,** Resolvent, Heat Kernel, and Torsion under Degeneration to Fibered Cusps, 2021

1313 **Paul Godin,** The 2D Compressible Euler Equations in Bounded Impermeable Domains with Corners, 2021

1312 **Patrick Delorme, Pascale Harinck, and Yiannis Sakellaridis,** Paley-Wiener Theorems for a p-Adic Spherical Variety, 2021

1311 **Lyudmila Korobenko, Cristian Rios, Eric Sawyer, and Ruipeng Shen,** Local Boundedness, Maximum Principles, and Continuity of Solutions to Infinitely Degenerate Elliptic Equations with Rough Coefficients, 2021

1310 **Hiroshi Iritani, Todor Milanov, Yongbin Ruan, and Yefeng Shen,** Gromov-Witten Theory of Quotients of Fermat Calabi-Yau Varieties, 2021

1309 **Jérémie Chalopin, Victor Chepoi, Hiroshi Hirai, and Damian Osajda,** Weakly Modular Graphs and Nonpositive Curvature, 2020

1308 **Christopher L. Douglas, Christopher Schommer-Pries, and Noah Snyder,** Dualizable Tensor Categories, 2020

1307 **Adam R. Thomas,** The Irreducible Subgroups of Exceptional Algebraic Groups, 2020

1306 **Kazuyuki Hatada,** Hecke Operators and Systems of Eigenvalues on Siegel Cusp Forms, 2020

1305 **Bogdan Ion and Siddhartha Sahi,** Double Affine Hecke Algebras and Congruence Groups, 2020

1304 **Matthias Fischmann, Andreas Juhl, and Petr Somberg,** Conformal Symmetry Breaking Differential Operators on Differential Forms, 2020

1303 **Zhi Qi,** Theory of Fundamental Bessel Functions of High Rank, 2020

1302 **Paul M. N. Feehan and Manousos Maridakis,** Łojasiewicz-Simon Gradient Inequalities for Coupled Yang-Mills Energy Functionals, 2020

1301 **Joachim Krieger,** On Stability of Type II Blow Up for the Critical Nonlinear Wave Equation in $\mathbb{R}^{3+1}$, 2020

1300 **Camille Male,** Traffic Distributions and Independence: Permutation Invariant Random Matrices and the Three Notions of Independence, 2020

For a complete list of titles in this series, visit the
AMS Bookstore at **www.ams.org/bookstore/memoseries/**.